Y0-DJQ-892

LET'S FIND
OUT ABOUT

BEES

LET'S FIND OUT ABOUT BEES

BY CATHLEEN FITZGERALD
Pictures by ARABELLE WHEATLEY

FRANKLIN WATTS | NEW YORK | LONDON

Library of Congress Cataloging in Publication Data

FitzGerald, Cathleen.
Let's find out about bees.

SUMMARY: Describes the busy life in a bee hive where queen, drone, and worker perform specific functions as contributing members of a tightly organized community.

1. Bees–Juvenile literature. [1. Bees] I. Wheatley, Arabelle, illus. II. Title.
QL568.A6F53 595.7'99 78-186938
ISBN 0-531-00079-6

Printed in the United States of America
3 4 5

LET'S FIND
OUT ABOUT

BEES

Bzzz — bzzz-bzzz.
Something flies by you in the garden.
It makes a loud buzzing noise.
What is it? It is a honeybee.

A honeybee has
5 eyes,
4 wings,
3 pairs of legs,
and 2 feelers.
Most bees also have a stinger.
Honeybees live on honey.
They make so much of this delicious sticky stuff that there is plenty left over for you and me to eat.

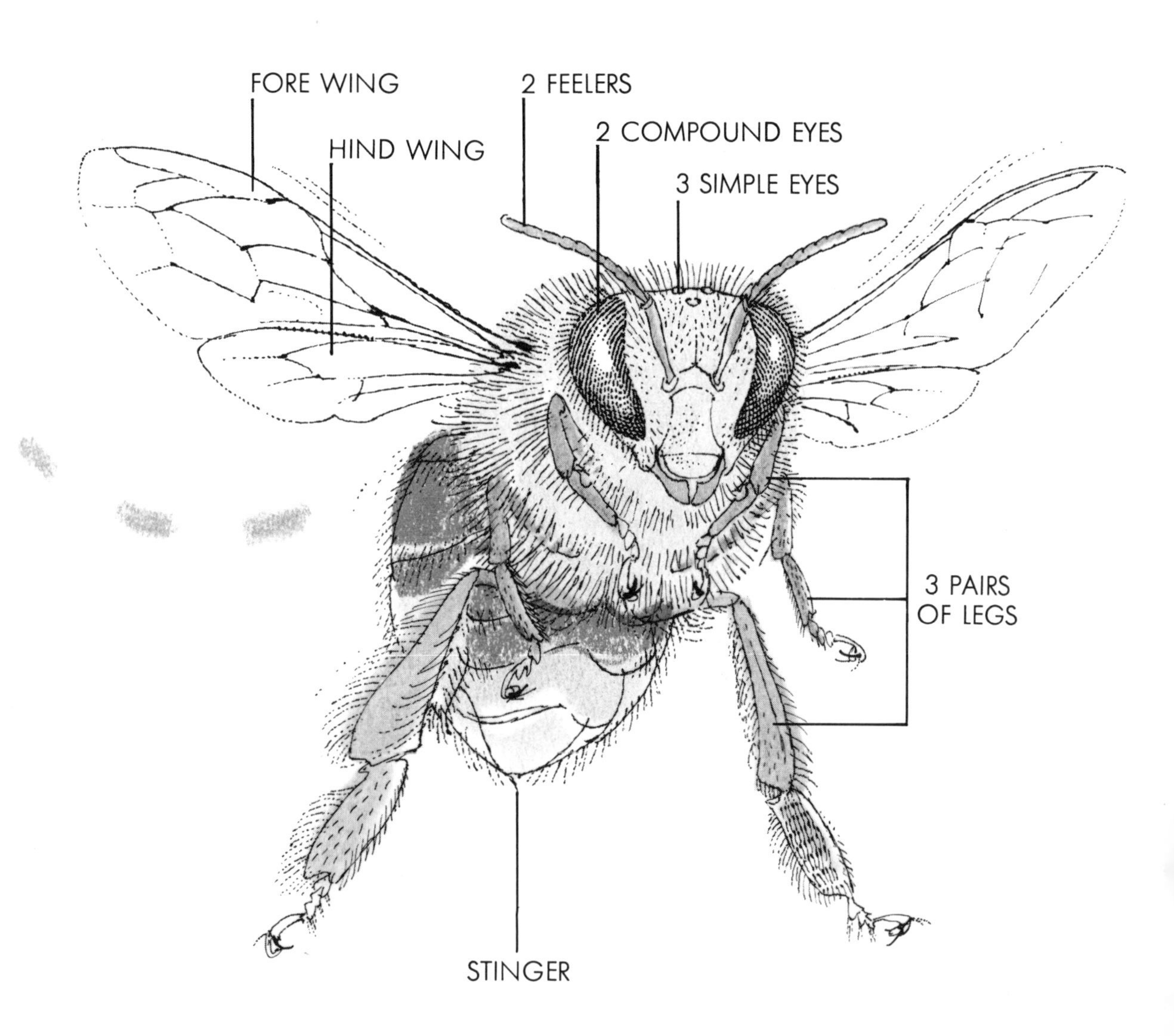
FORE WING
HIND WING
2 FEELERS
2 COMPOUND EYES
3 SIMPLE EYES
3 PAIRS
OF LEGS
STINGER

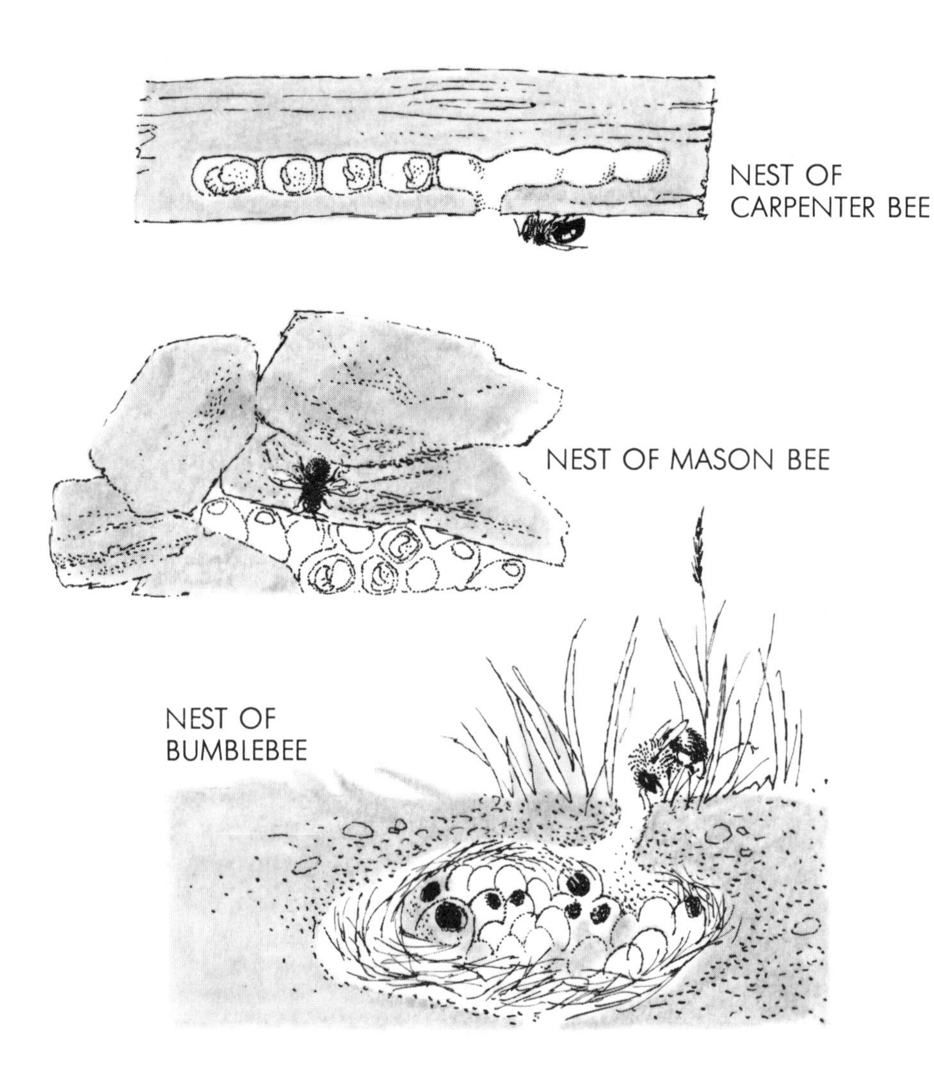

Most bees live in nests.

They build these nests in hollow trees or among rocks or in the ground.

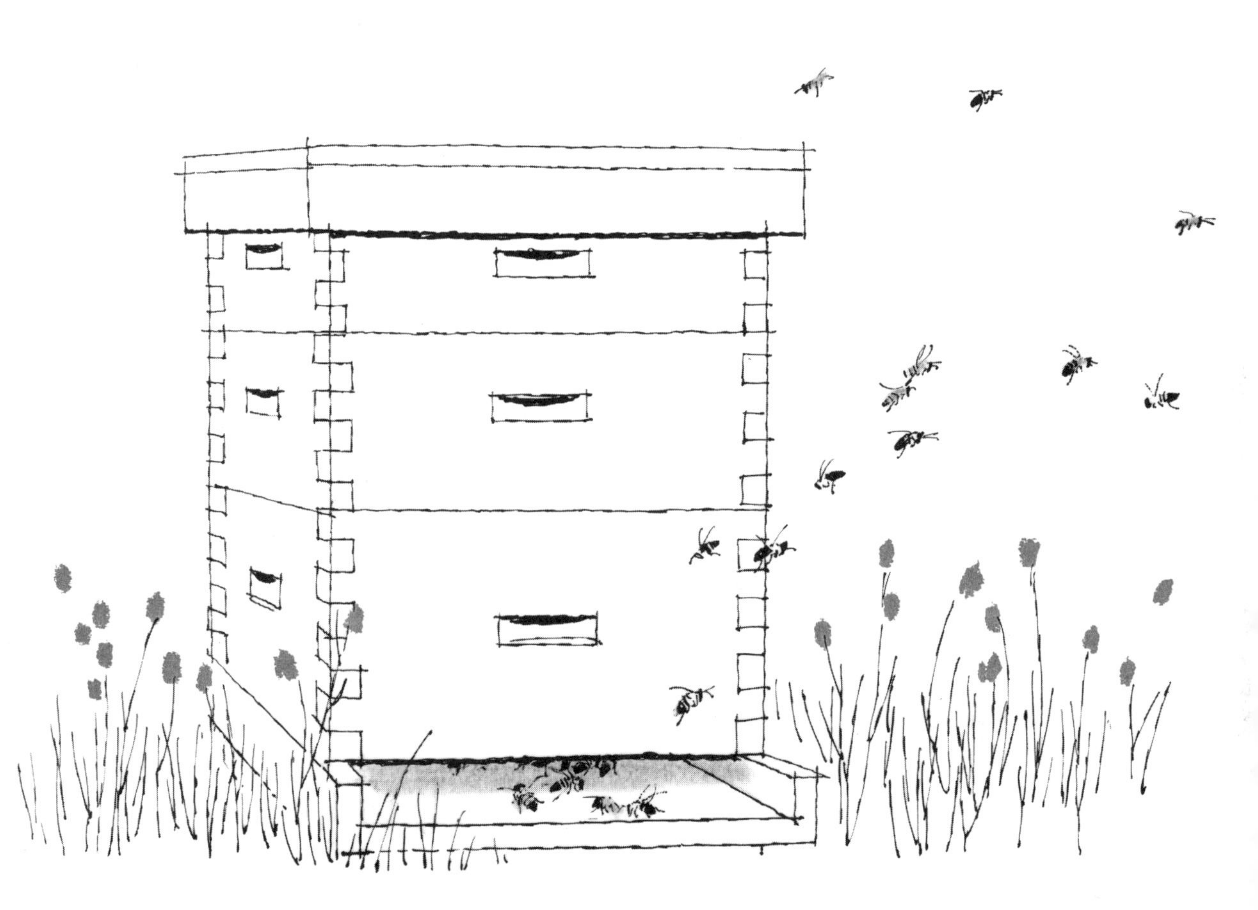

But many honeybees live in a wooden hive like
the one you see here.
This hive was built by a man called a beekeeper.

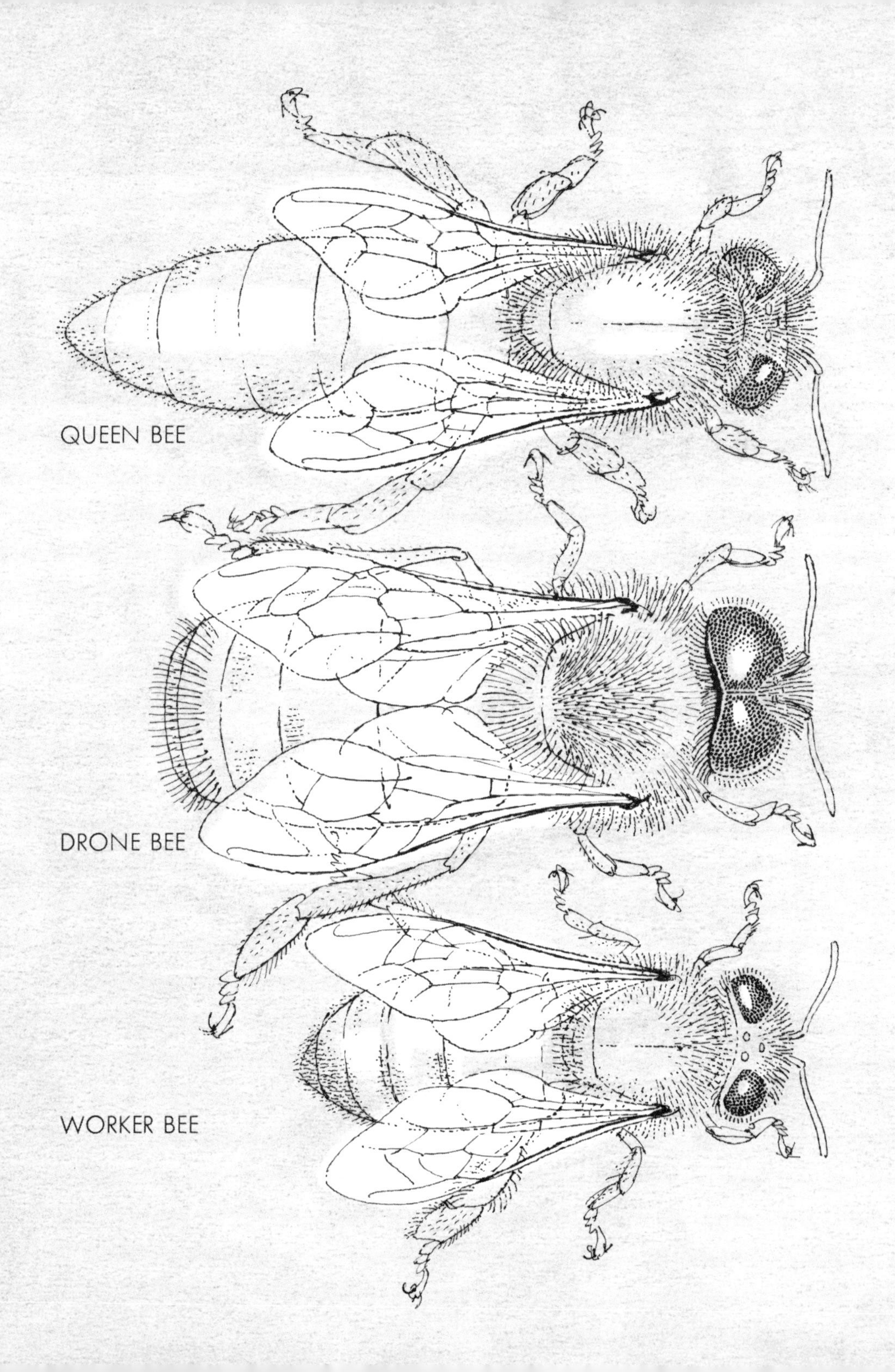
QUEEN BEE
DRONE BEE
WORKER BEE

In every hive there are three kinds of bees.
There is a QUEEN,
some DRONES,
and many, many WORKERS.
The queen does not look quite like the other bees.
Her body is bigger and longer and she is shinier.
She is the mother of all the bees.
There are thousands and thousands of bees in a hive, but there is usually only one queen.

Her most important duty is to lay eggs.
Baby bees come from these eggs.
Sometimes the queen lays as many as two
thousand eggs in one day.
Worker bees take care of the queen.
They stand around her in a ring and feed her

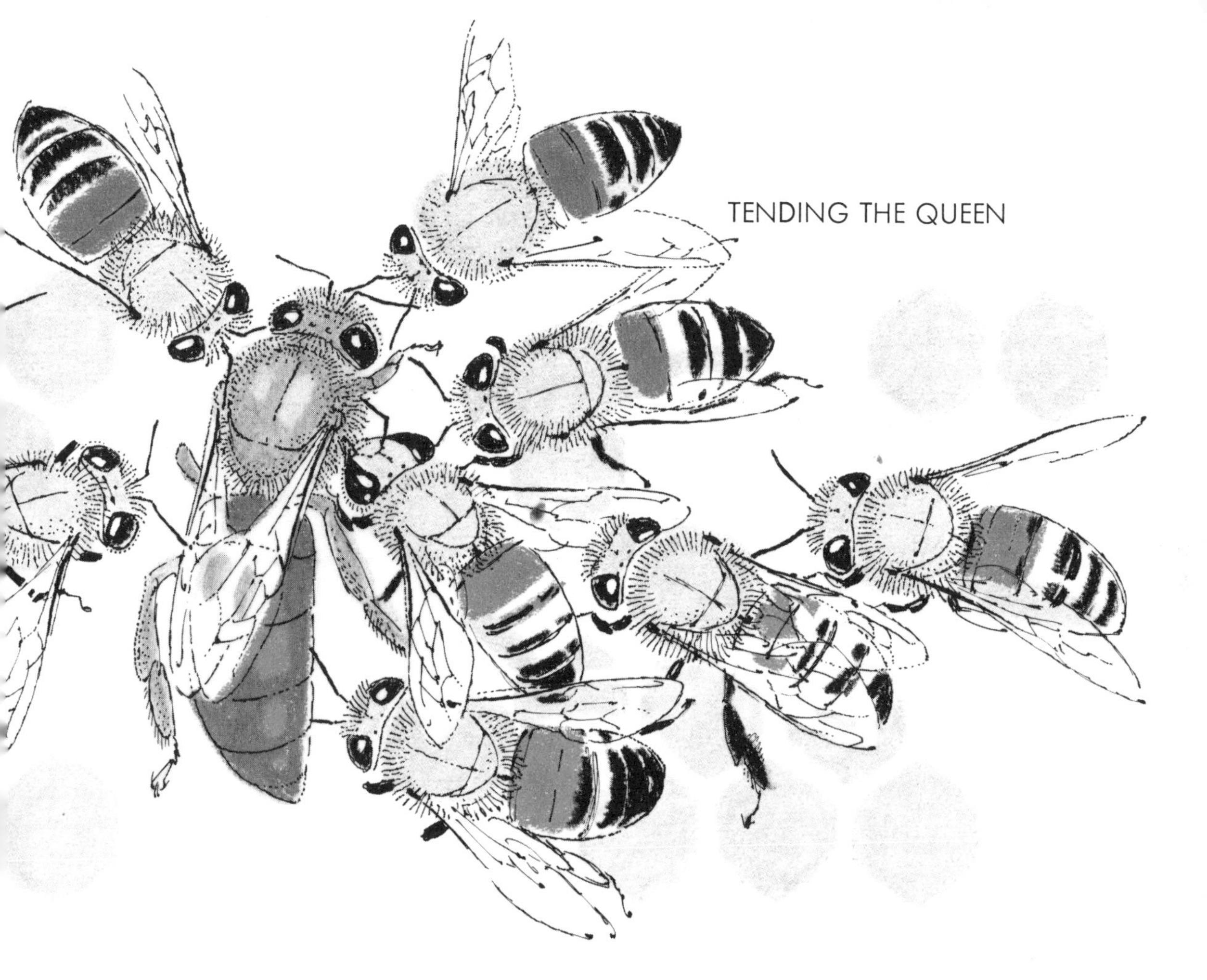

honey from their mouths.
They wash her with their tongues.
They even comb the tiny hairs on her body with
the hairs on their legs.
The queen does not have to do a thing for
herself.

The drones are male bees.
They are smaller than the queen, but bigger than the worker bees.
They have no stingers.
The drones do no work at all.
Their only job is to mate with the queen.
When the time comes for the queen to mate, she flies out of the hive and all the drones fly after her.
One of the drones, usually the fastest and strongest, mates with her.
This drone becomes the father of the new bees.

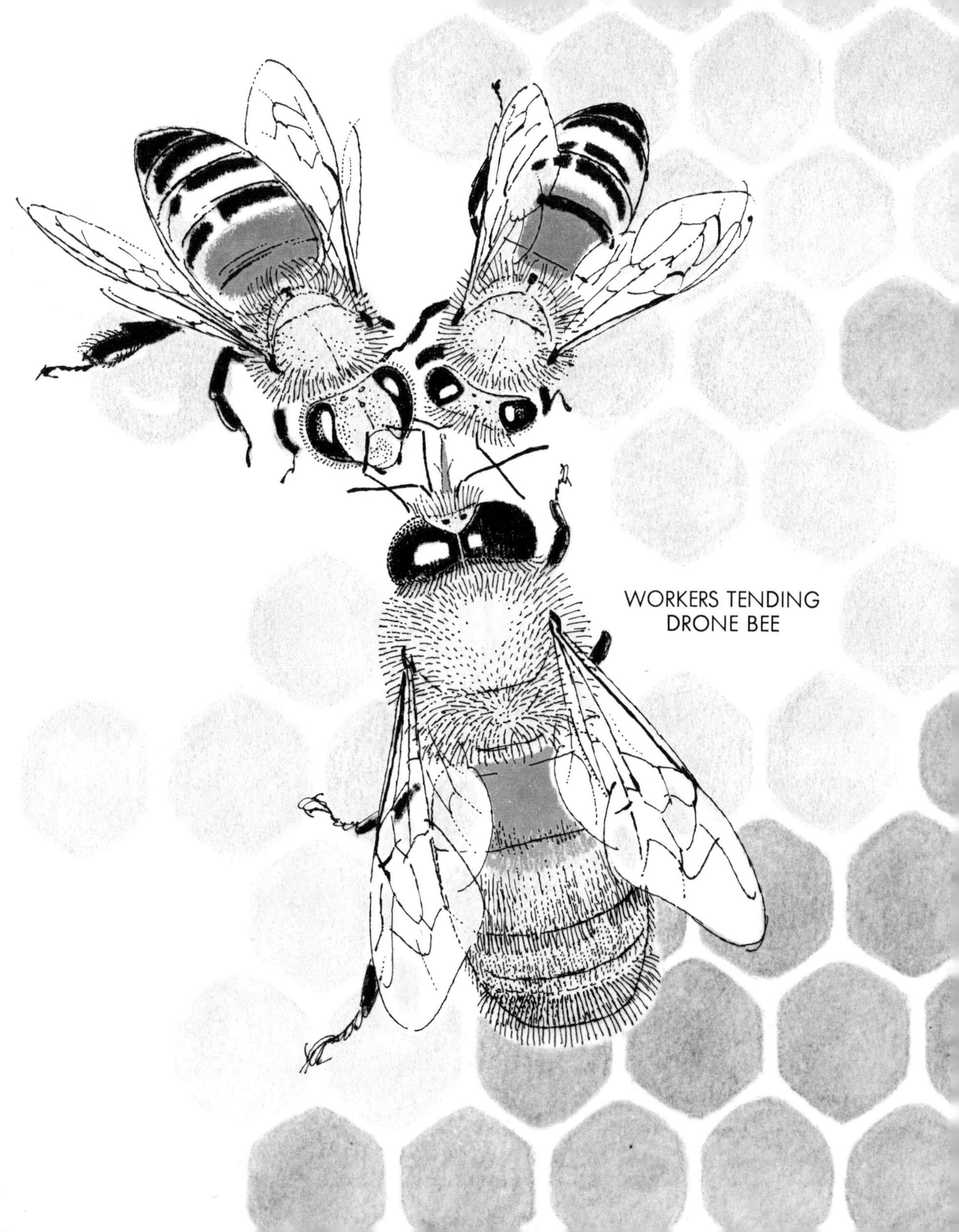

WORKERS TENDING
DRONE BEE

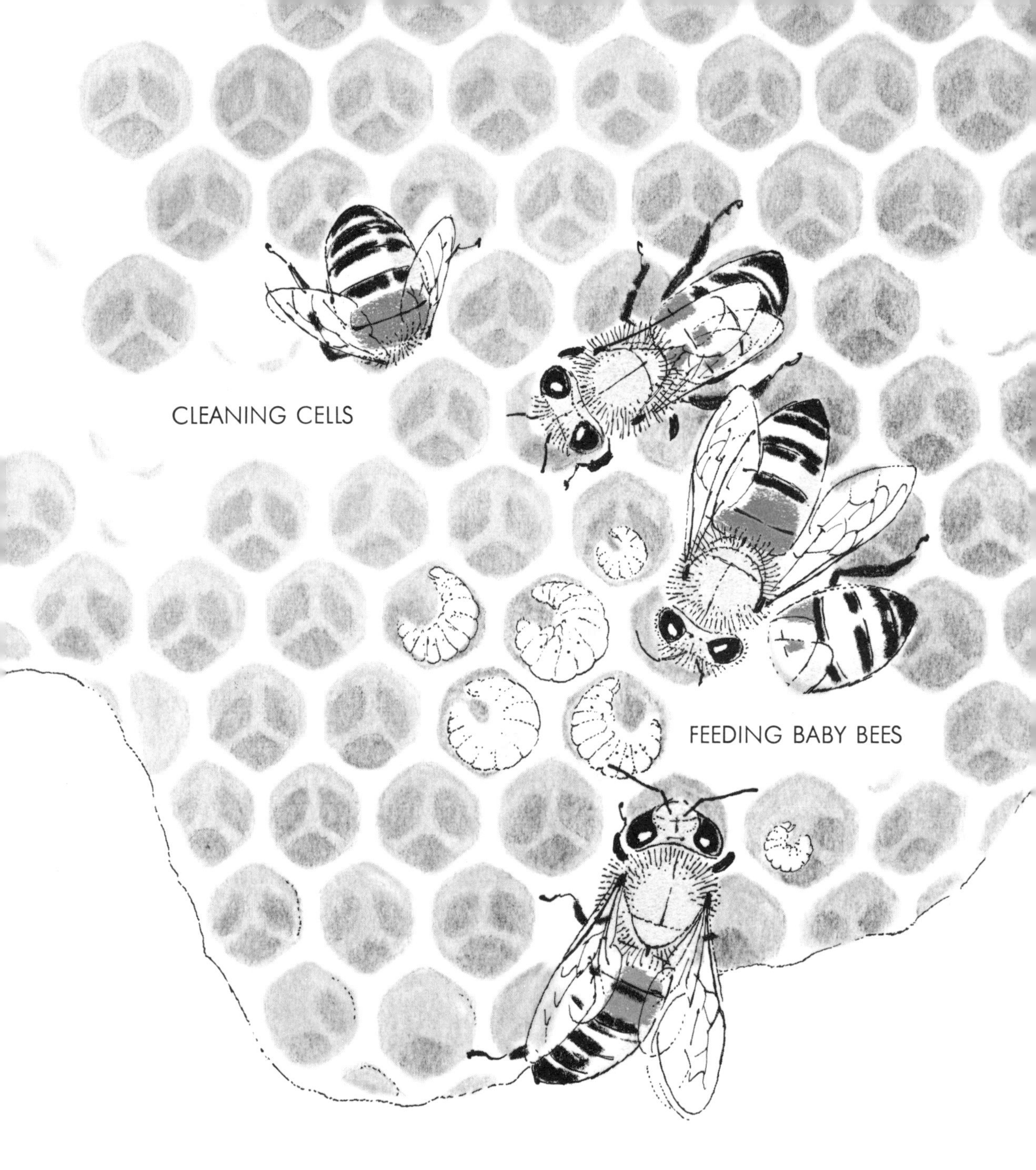
CLEANING CELLS
FEEDING BABY BEES

SECRETING WAX

Worker bees do all the work inside and outside
the hive.
They have many different jobs to do.
First, they are housekeepers who keep the hive
spotlessly clean.
Then they become nurses who look after the
baby bees and feed them.
Later on, they become builders.
They build the honeycomb with wax from
their bodies.

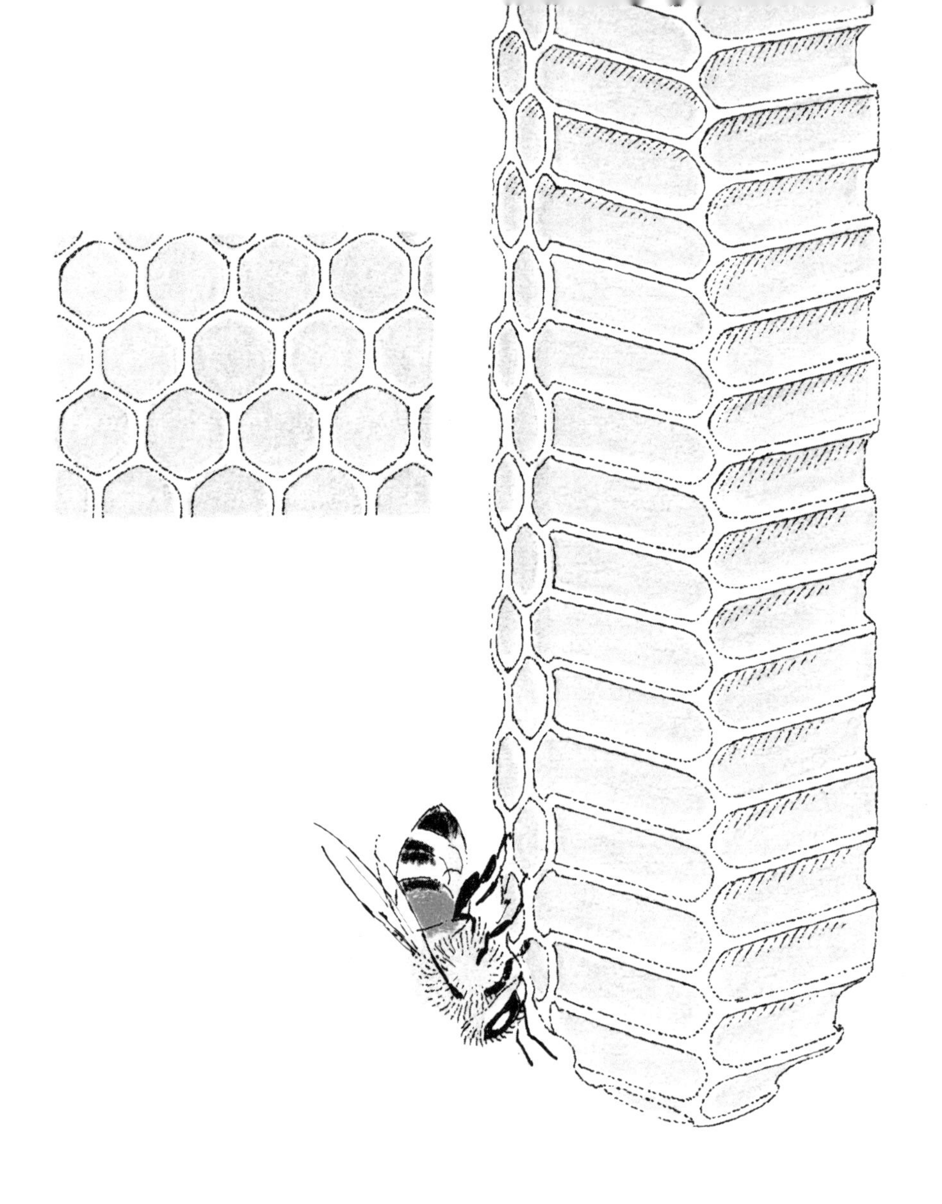

The honeycomb has thousands of little cells.
Each cell has six sides.

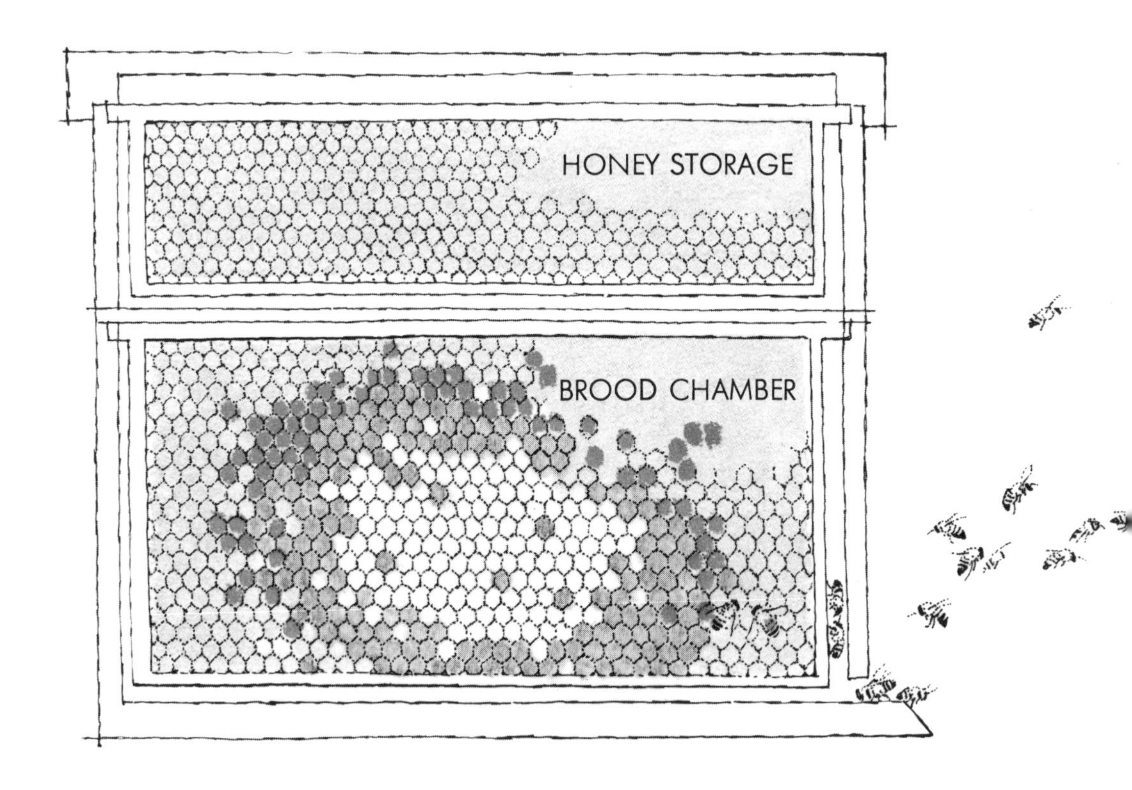

Most of the cells are used as storage places
for honey.
Some cells are used by the queen.
She lays her eggs in them.

Worker bees do some other jobs inside the hive.
Some of them receive food brought in by other
bees, and store it in the cells.
Some of them work as guards who stand at the
door and examine all the bees who fly in.
They know by the smell if a bee is a robber bee
or a stranger and they do not allow it
to enter.
Other bees are in charge of air conditioning.
The air-conditioning bees stand at the door and
fan their wings to keep the hive cool.

ENTRANCE TO HIVE

The worker bee's most important job is
making honey.
This is how she does it.

She flies from flower to flower.
Inside each flower is a sweet juice called nectar.
The bee sucks this sweet juice out with her
tongue, which is like a tube.
She stores the nectar in her body.
Then she flies back to the hive.

She gives the nectar to a house bee.
The house bee adds something from her body
that will turn the nectar into honey.
Then she puts the nectar in a cell.
When the nectar ripens, it becomes honey.

STORING
HONEY
CAPPING
FILLED
CELLS
GIVING NECTAR
TO HOUSE BEES

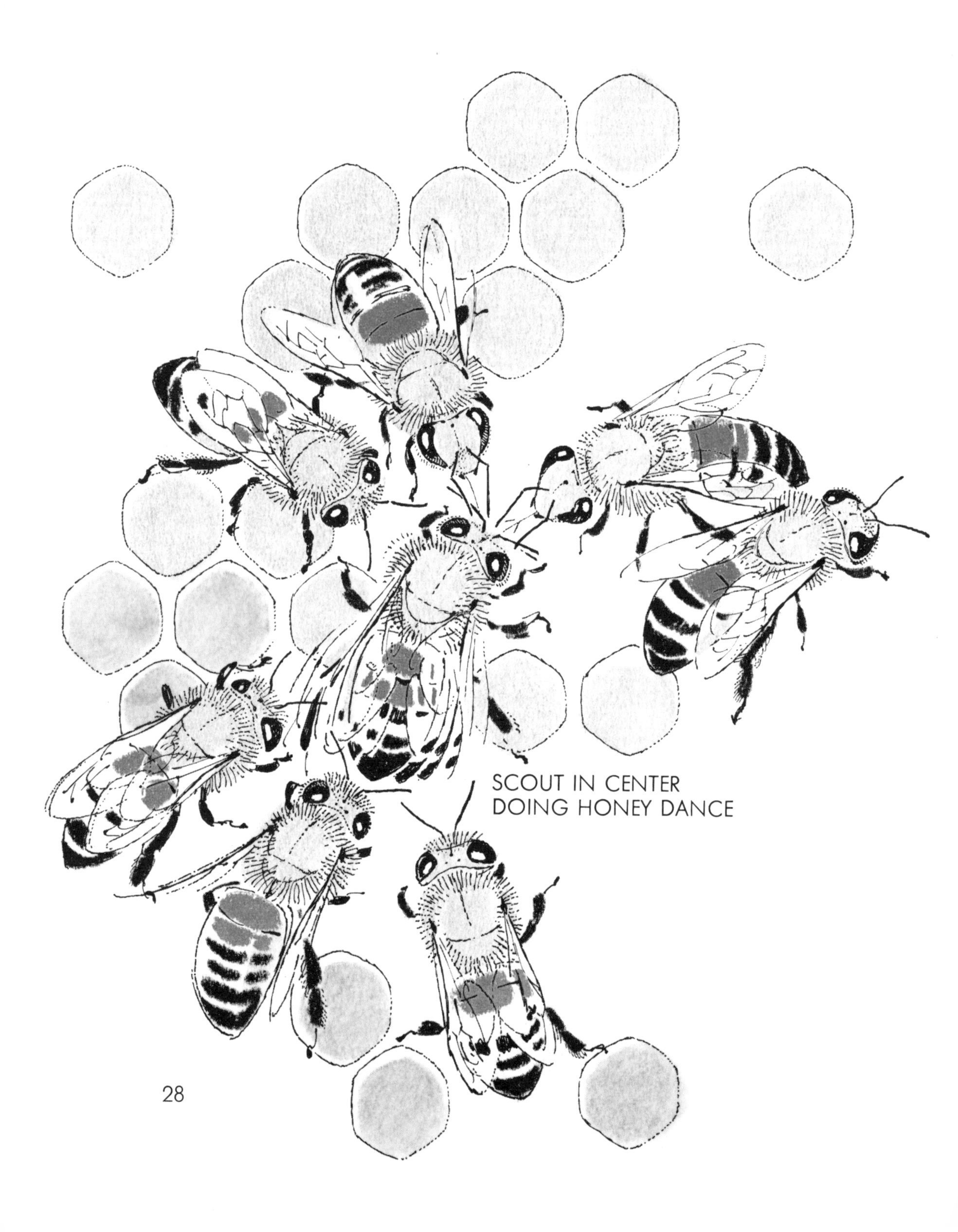
SCOUT IN CENTER
DOING HONEY DANCE

Some workers are scouts who look for nectar.
When a scout finds a good supply, she zooms back to the hive.
She tells the other bees where the nectar is by doing a special dance.
A round dance means the nectar is nearby.
A wagging dance means the nectar is farther away.
The other bees watch her carefully, then fly off in the direction of the nectar.

Bees also collect pollen.
Pollen is a kind of yellow dust inside a flower.
A worker bee carries the pollen on her hind legs, in flat places fringed by hairs.
These places are called pollen baskets.
If you see a bee flying back to the hive, you may notice the pollen on her legs.
She looks as if she were wearing furry yellow stockings.

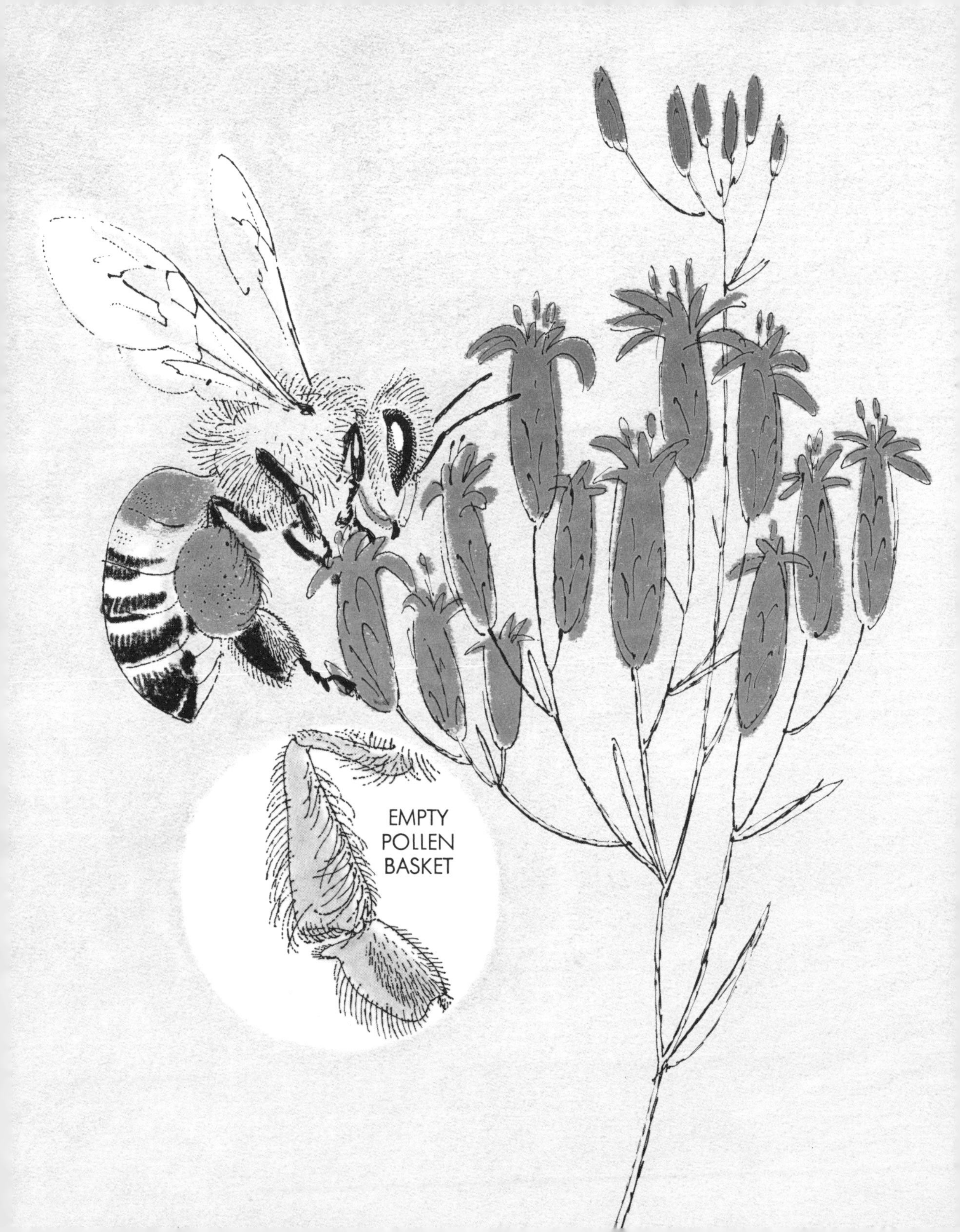
EMPTY
POLLEN
BASKET

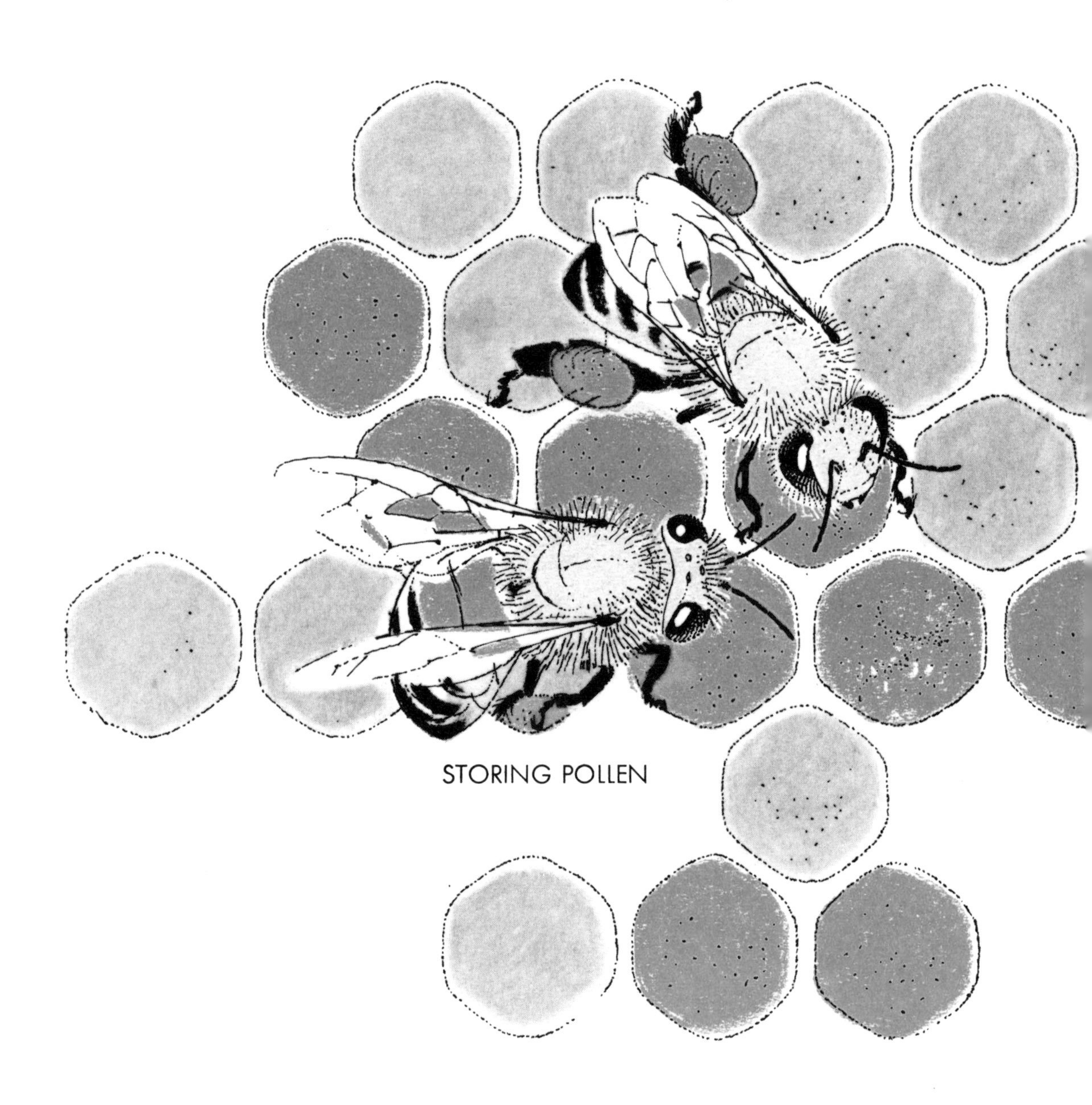

STORING POLLEN

Back at the hive, the bee puts the pollen in a cell.
She mashes it down and adds a little honey.
The mixture she makes is called bee bread.
Bee bread is the food that baby bees eat.

It is interesting to learn how a baby bee grows.
First, the queen lays a tiny pearly-white egg in a cell.
The egg is so very tiny that you can hardly see it.
After three days the egg hatches and a grub crawls out.
A grub looks like a very small worm.
For two or three days the nurse bees feed the grub with a special food called royal jelly.
Royal jelly is made in the heads of young worker bees.

QUEEN
LAYING
EGGS
WORKER
DEPOSITING FOOD

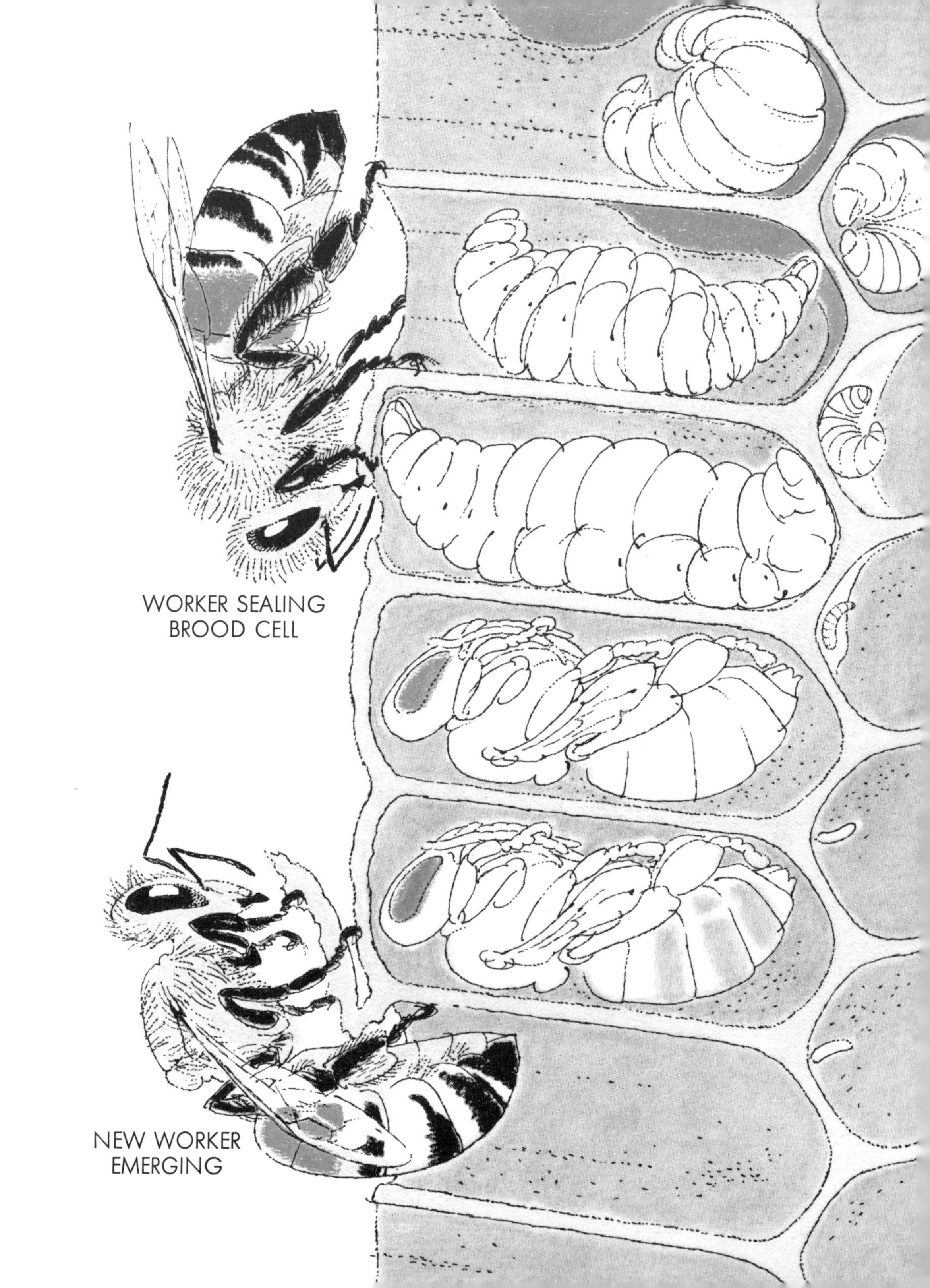
WORKER SEALING
BROOD CELL
NEW WORKER
EMERGING

For the next few days the grub is fed with
bee bread.
The grub grows and grows until it is nearly as
big as the cell it is in.
Then the nurse bees put a wax lid on the cell.
After twelve more days, the young bee breaks
the wax lid and out it comes, a fully
grown bee.

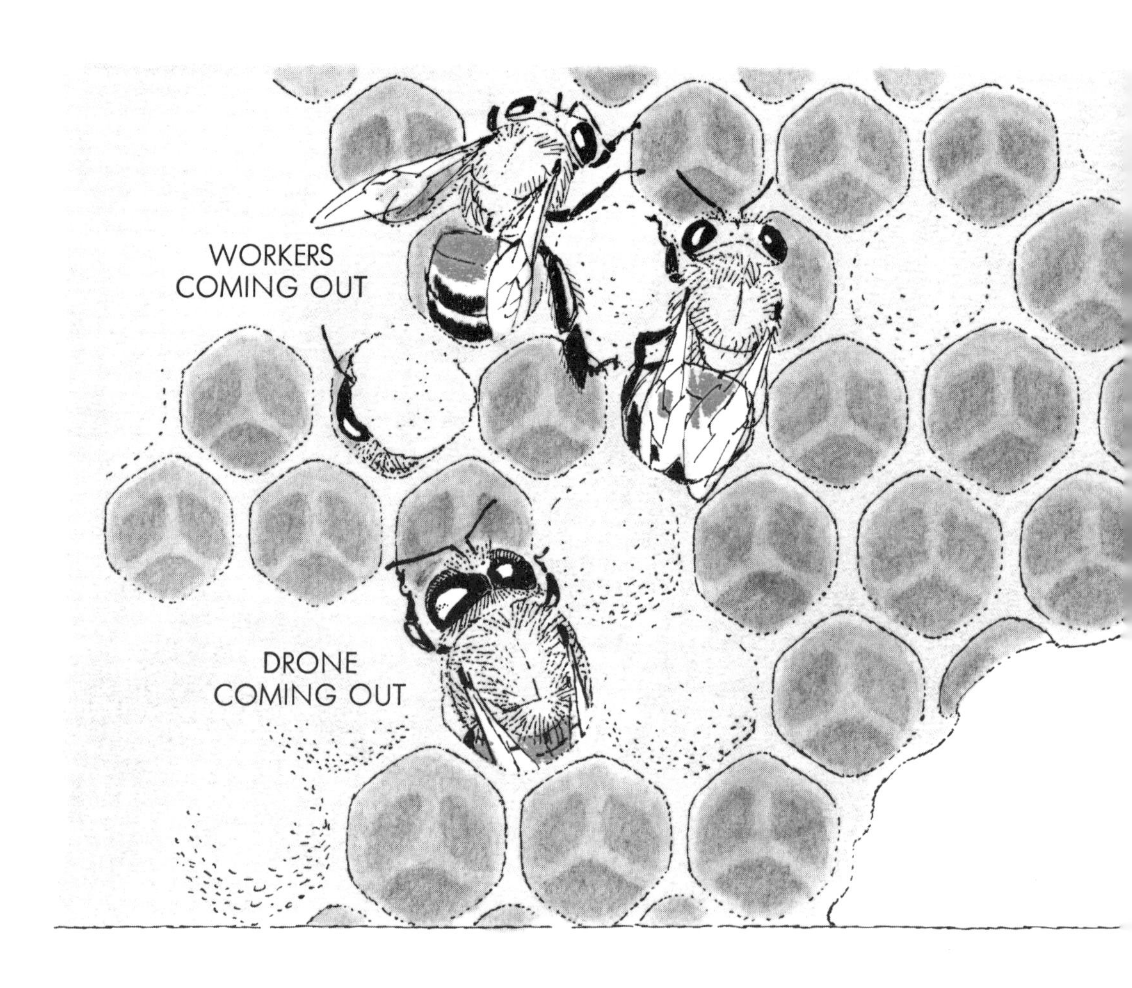

A worker takes twenty-one days to grow from
egg to bee.
A drone takes about twenty-four days.
But a queen takes only sixteen days.

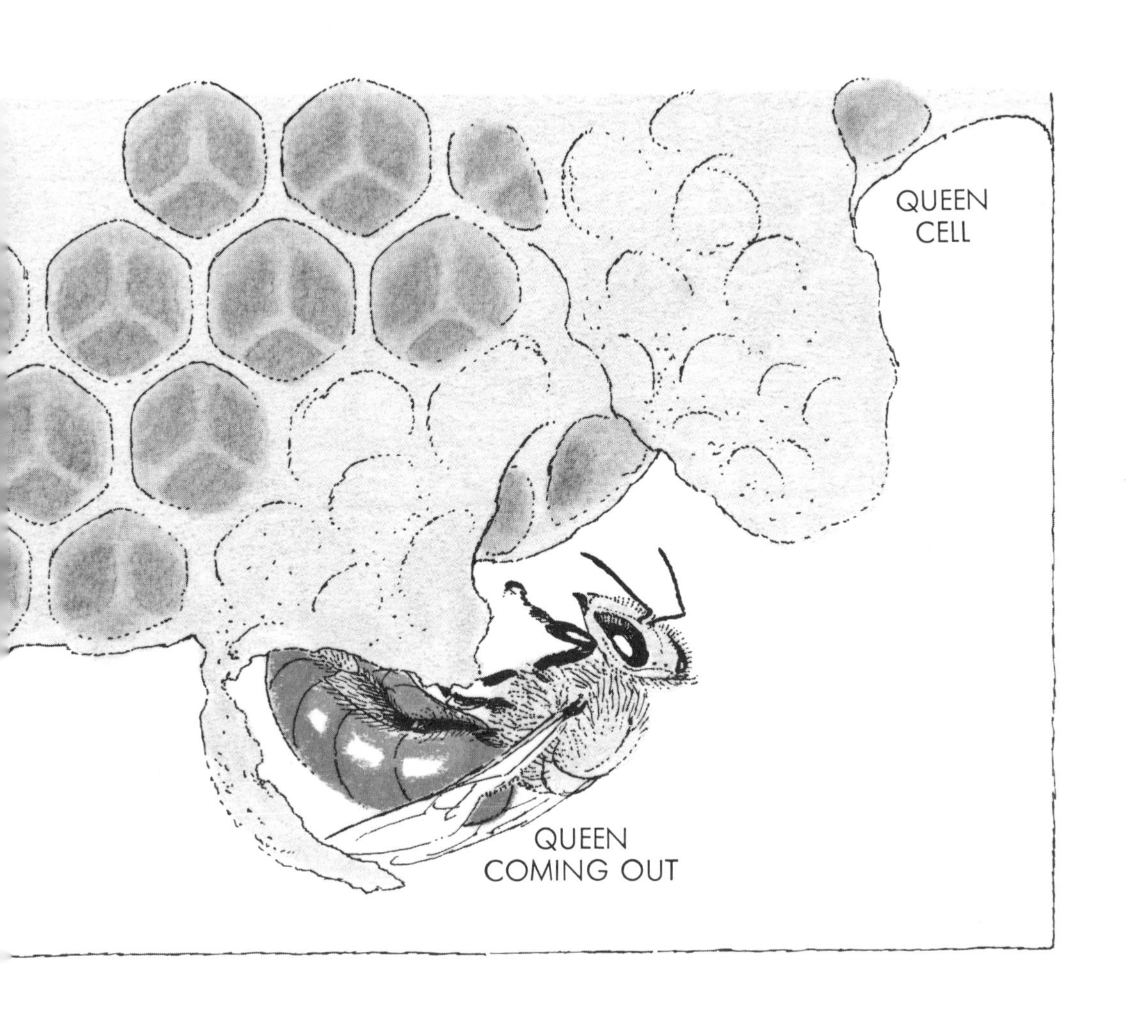

Queen grubs are not fed on bee bread at all,
but are given only royal jelly.
This food may be the reason they grow faster
than the other bees.

A beekeeper is a man who keeps bees and sells their extra honey.
When the beekeeper goes to the hive, he usually wears a bee veil so that he will not be stung.
First, he blows smoke into the hive.
This smoke seems to keep the bees quiet.
They continue to do their jobs and they pay no attention to the beekeeper.
He collects the honey.
Usually he strains the beeswax out of it.
Then he takes the honey to market.

There are many different kinds of honey.
Some honey is dark brown and some is golden
and some is yellow.
Every kind of honey has its own special taste.
The color and taste depend on the flowers that
the nectar came from.

CLOVER
HONEY
ORANGE BLOSSOM
HONEY
BLUEBERRY
HONEY

Where do bees go in winter?
You never see them around.
They seal themselves into the hive.
They live on the honey they made during
the summer.

Bees are important.
They make honey and wax, and they also do another big job.
They carry pollen from one flower to another.
A flower needs pollen from a different flower to make seeds.
Without seeds, new plants could not grow.
Bees are tiny creatures, yet they do a great deal of work.
They are always busy—as busy as a bee!

ABOUT THE AUTHOR

Cathleen FitzGerald was born in Dublin, Ireland. She is a graduate of the National University of Ireland and has a higher diploma in education from the same university. Second oldest in a large family, she started writing children's stories in her teens, to amuse her younger brothers and sisters. She has written scripts for television and radio and has broadcast her own work on Irish Radio. She now lives in New York City, where she is managing editor of the *New Book of Knowledge*.